50 Things to Know

50 THINGS TO KNOW ABOUT MUSHROOM HUNTING

Finding Edible Treasures in Your Backyard

Rebecca Wescott

50 Things to Know About Mushroom Hunting Copyright ©
2021 by CZYK Publishing LLC.
All Rights Reserved.

All rights reserved. No part of this book may be reproduced in any form or by any electronic or mechanical means including information storage and retrieval systems, without permission in writing from the author. The only exception is by a reviewer, who may quote short excerpts in a review.
The statements in this book are of the authors and may not be the views of CZYK Publishing or 50 Things to Know.

Cover designed by: Ivana Stamenkovic
Cover Image: https://pixabay.com/photos/mushrooms-forest-mushroom-forest-4944411/

CZYK Publishing Since 2011.

50 Things to Know

Lock Haven, PA
All rights reserved.
ISBN: 9798594902442

50 THINGS TO KNOW ABOUT MUSHROOM HUNTING

BOOK DESCRIPTION

Have you ever looked at a yard full of mushrooms that popped up overnight and wondered if they were poisonous? Have you taken a walk in the woods and stumbled on a mushroom you've never seen before? Does the idea of tasting amazing new flavors you found yourself appeal to you?

If you answered yes to any of these questions, then this book is for you...

50 Things About Mushroom Hunting, by author Rebecca Wescott offers an approach to understanding, locating, identifying, and eating these delicious wild fungi. While other books focus on identification or complex science, this book will recommend some quality identification guides and focus on the mystery and enjoyment of pursuing elusive mushroom surprises.

With this book, you will be off into the woods with your boots and bug spray, ready to start your mushroom-hunting experience.

By the time you finish this book, you will know what mushrooms are, where they grow, which ones are edible, and how to enjoy them. So grab YOUR copy today. You'll be glad you did.

TABLE OF CONTENTS

50 Things to Know
Book Series
Reviews from Readers
BOOK DESCRIPTION
TABLE OF CONTENTS
DEDICATION
ABOUT THE AUTHOR
INTRODUCTION
1. What Mushrooms Are and What They Do
2. Recommended Guides for Finding and Identifying Mushrooms
3. Hunting Mushrooms: Worth it or Not?
4. Using the Environment to Find Mushrooms
5. Seasonal Mushrooms and their Timing
6. Factors That Affect Mushrooms Coming Up
7. Basic Types of Mushrooms and What to Do with Them
8. Chicken of The Woods
9. Chanterelles and Black Trumpets
10. Lion's Mane
11. Maitake or Hen of the Woods
12. Oyster Mushrooms
13. A Few Mushrooms NOT to Eat
14. Identification Tricks: Gills

15. Identification Tricks: Spore Prints
16. Helpful Tools to Take in The Woods
17. My Personal Favorite Mushroom Hunting Tool
18. How to Look
19. Where to Look (Including Public Property)
20. Tips for Preparing Your Mushrooms at Home
21. Things You Will Learn to Put Up With if You Eat Wild Mushrooms
22. Basic Cooking Methods
23. Tips for Cooking with Wild Mushrooms
24. Giving Your First Meal a Test Run
25. Possible Mushroom Side Effects
26. Storing Mushrooms at Home
27. Who Shouldn't Eat Wild Mushrooms?
28. Can Wild Mushrooms Kill You, and How?
29. Dangerous Wild Mushrooms and How to Spot Them
30. Look-Alike, Not Taste Alike
31. Do People Collect Non-Edible Wild Mushrooms?
32. What Health Benefits Might Mushrooms Have?
33. Where Can You Find Medicinal Mushrooms?
34. How Do Other Cultures Use Wild Mushrooms?
35. Can You Harvest Mushrooms and Make Your Own Medicine?
36. How Long Have People Been Harvesting Mushrooms?

37. Mushroom Hunting: Old Methods Work
38. Most Important Skills to Learn for Mushroom Hunting
39. Never Guess About These Things with a Mushroom
40. Reasons to Take Kids Mushroom Hunting with You
41. Hunting Mushrooms and Respecting Nature
42. Why We're Not Talking About Hallucinogens in this Guide
43. One Mushroom to Avoid Unless You Want to Take a "Trip"
44. Mushroom Hunting Competition to Watch Out For
45. Signs There Might Be Mushrooms Around
46. Am I Hurting the Fungus if I Pick All the Mushrooms?
47. Why has mushroom hunting become more popular?
48. Local Clubs and Groups
49. Why People Think Mushrooms are Bad
50. What Mushrooms Can Teach Us About the World Around Us

HELPFUL RESOURCES:
50 Things to Know

DEDICATION

Dedicated to my husband, Steve, who has helped me learn how to stop getting lost in the woods quite so often and encouraged my writing every step of the way. Also, to Steve's brother Tommy, who shares my enjoyment.

ABOUT THE AUTHOR

Rebecca Wescott received her first glowing praise for her high school work and has been hooked ever since. After a BA in English and Creative Writing from Kenyon College, she completed a Master's in Education (Science and English) at the Ohio State University. For the past three years, she worked as a content manager, writer, and editor for a heavy industrial machine company.

She has worked as a teacher, behavior specialist for children with mental health issues, a drug and alcohol counselor, an English teacher, a gas station attendant, a salesperson for industrial equipment, and a content writer and manager for a heavy manufacturing company.

Nature walks, identifying plants and fungi, and harvesting mushrooms to eat have been hobbies for many years, including membership in local and national mushroom hunting clubs. The author spends her warmer days walking at the nearby lake, staring at the ground, and turning things over with sticks.

The author has plans to become a crazy cat lady in the future and feels that five cats are a good start. She collects pens, plays a vicious game of trivial pursuit, and has spent many years developing strategies to prank her teenage daughter in increasingly bizarre ways.

INTRODUCTION

Mushroom experts, formally called mycologists, have a strange sense of humor. Curtis Gates Lloyd, an expert in the field, left this monument as his tombstone:

> *"Monument erected by himself, for himself, during his own lifetime to gratify his own vanity. What fools these mortals be."*
>
> \- Curtis Gates Lloyd

Store-bought mushrooms taste bland. Wild mushrooms come with their own exciting flavors, from delicate and fruity like chanterelles to deliciously meaty like chicken of the woods. A few wild mushrooms, like oyster mushrooms and lion's mane, can now be grown commercially.

Most wild mushrooms, though, do not grow unless their specific needs are met. If you want them, you will have to either buy wild-harvested ones (very pricey) or harvest your own with just a good pair of boots, a reliable field guide, and a bit of luck. I have a few preferred tools I always take with me, but we'll review equipment later.

Please do not forget that mushroom hunting can be dangerous, and eating the wrong mushrooms can be fatal. There is no substitute for an experienced mushroom hunting partner and a quality field guide for your area.

1. WHAT MUSHROOMS ARE AND WHAT THEY DO

Mushrooms are the spore-producing bodies of a fungus. Most of the fungus, called a mycelium, lives inside the material it grows on. Pull a piece of crumbling wood off a stump, and you may find white threads running through it. These are hyphae, the strands of fungus that make up the mycelium. Most of the action of digestion and growth happens here. Mushrooms and other odd fruiting bodies are made by the fungus to spread its spores around.

Fungi have more in common with animals than plants. They can't produce their own food with chlorophyll and sunshine. They digest organic matter. Sometimes it could be the bread on your counter, and sometimes it could be the tree that falls on your car. One mushroom or bracket fungus can produce billions of spores.

When you see a mushroom, you are not seeing the whole fungus. It is more like you are only seeing the apples and the rest of the apple tree is underground. When we harvest mushrooms, we do not harm the fungus. It can always make more mushrooms if we respect the environment it needs to grow.

2. RECOMMENDED GUIDES FOR FINDING AND IDENTIFYING MUSHROOMS

Get a good guidebook, and then get a few more. A specimen can look much different newly emerged, and after three days of slugs munching on it. Each area of the world will have local guidebooks for that area, and I encourage you to get at least one of these to go with the more general ones.

Remember that no guidebook will ever substitute for the advice of a knowledgeable guide, and if you can seek out a mushroom group in your area, take advantage of it.

The most popular guidebook to mushrooms is "National Audubon Society Field Guide to North American Mushrooms." This exhaustive resource is dense but thorough. Another amazing resource is "The Complete Mushroom Hunter, Revised: Illustrated Guide to Foraging, Harvesting, and Enjoying Wild Mushrooms" by Gary Lincoff, a guru of mushroom hunting in the eastern USA. A third resource, "Mushrooming without Fear: The Beginner's Guide to Collecting Safe and Delicious Mushrooms" by Alexander Schwab, offers basic and sometimes over-simplified advice, but you will not eat anything too awful if you follow its guidelines.

Flip through each and read about the types of mushrooms you think you might want to eat. Identification is the key to hunting wild mushrooms for your meals without putting yourself in danger.

3. HUNTING MUSHROOMS: WORTH IT OR NOT?

With numbers of wild mushroom hunters growing every year, more and more people seem to seek the unique flavors and the adventure (and occasional frustration) of the hunt. Since most wild mushrooms will not grow in captivity, finding them results from a combination of luck and experience.

Wild mushrooms can be worth money. During some years, prized morels go for exorbitant prices. Morels like places like old, neglected fruit orchards, but they also love areas of woods that have been in a forest fire. After some fires, the Pacific Northwest has erupted with such vast morel harvests that prices dropped through the floor. In other seasons, filling a basket can require hours of searching. Since the mushroom is just the fungus's fruiting body, during a bad year, it might not waste energy making mushrooms at all.

My personal favorite mushroom, chicken of the woods or sulfur shelf, cannot be relied on to come up anywhere a second time. This is the large fruiting body of a wood-decay fungus, and spotting the neon orange and yellow amidst the forest colors can send chills up your spine. If the price of wild mushrooms and the lure of their flavor doesn't draw you in, consider the thrill of the hunt and the adrenaline rush of finding your prey.

4. USING THE ENVIRONMENT TO FIND MUSHROOMS

When hunting wild mushrooms, habitat makes all the difference. Lion's mane and its delicious relatives only grow on wood, often high up a tree, so keep an eye out in an area with mature trees. Many other fungi have long-standing relationships with their partner trees and will only pop up in that vicinity. This includes the chanterelles, golden treats with a delicate flavor. My chanterelle spot reliably produces basket after basket of them each year. Still, it requires a hike to that exact spot to find them.

Around the base of dead or dying trees, you may find fungi that go by the charming name of "butt rot." These wood eaters create many of the hollowed-out trees found in most forests. Rather than forming a nice partnership, Honey mushrooms seem to pop up in clusters around a doomed tree. Trees surrounded by honey mushrooms, or with any mushroom growing on the trunk, should be inspected by a professional. The classic bright red with white spots mushroom, fly agaric, likes pine trees but causes no harm, like chanterelles and other symbiotes.

Sometimes the land itself affects where mushrooms grow. Black trumpets, a rich-flavored relative of the chanterelles, hide so well on the forest floor that I have missed them many times. Because they are paper-thin, you need a lot to add their earthy marvels to a meal, so finding a handful one day was a novelty until I realized there was another cluster a few steps down the slope of the hill, then another and another. The fungus had either slid down the hill as the soil moved, or water had carried its spores downhill. When mushroom hunting, keep your eyes on the forest floor.

5. SEASONAL MUSHROOMS AND THEIR TIMING

Mushrooms have seasons, just like plants do. A few fruiting bodies can last for years, but they are bracket fungi or polypores that form hard, woody shelves over many years of growth. The birch polypore, which starts out as a small white lump on bark, expands into a tough shelf tinted green with algae growing on top. Like the medicinal mushroom known as reishi, others form shiny, flexible burgundy shelves that harden into one you could set your water bottle on.

Some mushrooms have a notoriously short season for harvesting. Once a mushroom comes up from the soil, it gets attacked by bugs, slugs, bacteria, other fungi, and even squirrels and deer. Get them fast or get none. Morels are rumored to appear "when the oak leaves are as big as mouse ears," according to stories. I expect my chanterelles to start appearing after the first big rain in the second half of June. Maitake or hen of the woods clusters (also called sheepshead), strictly a fall mushroom, can weigh more than twenty pounds.

You can expect to start finding mushrooms, although not always edible ones, as soon as spring rain starts and the soil warms up. They flourish through the spring, summer, and well into the fall since the underground mycelium has protection from dryness and temperature changes. Practice identifying your non-edible ones early, so you have improved your skills when the good ones show up. You will learn a lot from finding out what grows seasonally in your area.

6. FACTORS THAT AFFECT MUSHROOMS COMING UP

Dampness does seem to wake up the production of mushrooms. However, attribute part of that to the fact that fungi need no light, so they can live in damp, dark places. A prolonged dry season can have a drastic effect on mushroom hunting. My reliable chanterelle patch, which had been arriving for years with clockwork precision, neglected a very dry June and waited to make their appearance in late July, nearly a month late.

Some mushroom spores don't need wind to spread them around. If you have ever seen anything in your yard that looked like a prop from a dirty movie and smelled like dead meat, you have met a stinkhorn. They rely on flies to show up for the smelly buffet and carry the spores away with them. Others, like the tiny but dramatic artillery fungus, need rain so they can absorb water. The water pressure will become so great it turns the entire cap inside out and flings the spores several feet away.

Study local weather patterns in your area, including when the ground usually freezes and thaws for the year and what months offer the most rain.

While rain brings out the mushrooms, hunting for them in a rainstorm may not be fun. Also, check weather forecasts because roaming a forest during a lightning storm is not recommended.

7. BASIC TYPES OF MUSHROOMS AND WHAT TO DO WITH THEM

Fungi come in several big families, but most of them do not make mushrooms. Technically, a "mushroom" is a general word for a gilled fungus. Still, most mushroom hunters apply it to any visible fungus fruiting body. If you pick this type of mushroom and flip it over, you will see that it has smooth gills radiating from the center. The spores develop on these gills and fall out.

Some excellent edible mushrooms are boletes. A bolete will have no gills when turned over. The bottom will usually look like a sponge with very tiny holes. You may struggle to identify boletes since many look similar. One technique for identification: press on the sponge surface with your finger or slice into the stem. Some boletes turn deep blue when damaged. Some will change immediately but keep watching for a minute or two.

Mushroom identification guides include boletes and gilled mushrooms because both are common, and some are edible. You cannot assume that any mushroom, bolete or gilled, is safe to eat without a solid identification. Always test a bolete to see if it bruises, and always dig around the base of a gilled mushroom instead of picking it (we will talk about that later). You can always collect mushrooms of questionable edibility and take them home for further inspection.

8. CHICKEN OF THE WOODS

This fluorescent polypore has sent me stomping through the trees, only to find out I had been chasing a hunter's old hat. Ranging from yellow-orange to vivid tangerine, these may be among the most distinctive mushrooms you can find. They have no gills and almost invisible pores. The surface looks puffy or lumpy, and the part closest to the tree gets very tough, but the rest of the mushroom can be eaten. If you can avoid it, do not break the shelf off at the base since this part is inedible and usually has bark and dirt all over it.

You can find this mushroom high up in a tree, growing near the base, or on dead trees on the ground. Several species have different preferences, but all are edible. For some people, chicken of the woods growing on conifers can upset the stomach, but this is why we test our mushrooms before starting the meal. This wood decay fungus likes already-damaged trees. Its appearance should tell you that a tree has been seriously compromised by fungus mycelium. If it is close to your house, you may want to consult a tree expert.

9. CHANTERELLES AND BLACK TRUMPETS

These fungi get their name from their shape. Chanterelles might be only indented in the center, but black trumpets look like little funnels. Chanterelles have false gills, and before going out to hunt for them, look up pictures of real and false gills to see the difference. The false gills on chanterelles run down the stem and have forks and connections between them. Getting a good idea of what this looks like will keep you away from this mushroom's only unpleasant look-alike.

Chanterelles share a bond with the roots of mature trees without being picky about the species of tree. The relationship benefits both, and the chanterelles do not eat their tree friend. Remember this when mushroom hunting because chanterelles never grow out of wood. They grow from underground, where the tree's root system lives. Anything similar but growing on wood is not a chanterelle.

Chanterelles have a light fruity smell, but you may not smell it till you have a basket in the back of your car. Black trumpets have an earthy, rich aroma. Make sure you harvest plenty because chanterelles contain a lot of water and will cook down to a fraction of their size. Black trumpets, thin as paper, make more of a seasoning than an addition to the meal. A jar of dehydrated and crumbled ones can be used to add that earthy flavor to meals for months.

In my area, chanterelles tend to be predictable in their arrival and harvest times. If you can find out when and where in your area a bath pops up, you will probably find it every year. Remember that a tree's root system can spread for yards, so a big patch of these beauties can be partners to the same tree.

10. LION'S MANE

These bizarre fungi look nothing like a mushroom. They look more like a cluster of icicles or a white beard. Members of this family can be found worldwide, so lion's mane will have relatives with many other names. Some countries consider them medicinal. They are all edible.

The icicle clusters of these fungi can grow on living trees or dead hardwoods on the forest floor. Locally, lion's mane seems to prefer live or dying beech trees. Still, many other species in another area will have a different favorite. These fungi are hard to misidentify but check out a picture of Northern Tooth if you are in North America. This fungus isn't a close look-alike, but it does grow on trees and have tooth-like spines. It will not hurt you, but it will not taste good.

This fungus can now be grown commercially, but people outside larger cities will find it unavailable or expensive. Its popularity in gourmet cooking comes from its texture and flavor, which many compare to crab or lobster. Mushroom hunters often chop these up and fry them like crab cakes. As a vegan substitute for leggy seafood, their popularity should continue to grow.

When bringing these home, make sure you break up the clusters right away. The nooks make great hiding places for bugs. Don't panic: most wild mushrooms have bugs in them, and you'll have to get used to getting rid of them.

11. MAITAKE OR HEN OF THE WOODS

Also known as Ram's Head or Sheep's head, this polypore lives in many parts of the world. In the USA, they prefer the northeastern part of the country. This large mushroom has "petals" or "feathers" all emerging from a central mass. These masses can weigh over twenty pounds. Most common in association with oaks, they also like other hardwood trees. As a polypore, it has tiny holes on the underside instead of gills. Like bird feathers, the color is often mottled grayish brown, but the sheer size will keep you from missing them.

In Asia, this fungus is known as maitake and reported to have a host of health benefits, curing everything from diabetes to arthritis to erectile dysfunction to cancer. Research continues, but most mushrooms contain interesting chemicals science has not yet explored.

    Hen of the woods may have a chalky flavor, especially as the mushroom cluster gets larger and older. The most pliable and soft pieces can be cut off and cooked but leave the hardened parts or those in the center. This is a late-season mushroom with a peak season in early autumn in my region of the world. They can reoccur in the same places for several years in a row, and some mushroom hunters I know have had to take wheelbarrows out to bring home a huge sample.

    The most common look-alikes for this are the black-staining polypore and Berkely's polypore. Separating the three is easy enough: the black-staining polypore will stain black if torn or cut open. The staining takes a few minutes, so give it time before you start harvesting. The Berkeley's polypore, also large, does not have the "feathers" but large shelves. A glance in a guidebook should separate them. Both look-alikes are harmless but not desirable for eating.

12. OYSTER MUSHROOMS

Beautiful and transient, oyster mushrooms pop up out of nowhere, but they are eaten by pests or other fungi just as fast. These wood decay fungi have gills and a stem, but the stem usually emerges from the side. Several species of oysters, all edible, can now be commercially grown and sold in stores. You can buy a kit and grow a batch of them at home.

Oyster mushrooms grow on dead wood and can fruit several times on the same dead tree. Using a guidebook should help you make sure you have oysters, but the first thing to check is whether there are gills. If the mushroom has pores or a smooth-looking surface, it is something besides an oyster.

While I like hunting for my own mushrooms, I cannot complain about oysters' availability in stores. When found fresh, these mushrooms have a delicious, mild flavor that suits almost anything. A day or two late, and you may find them being dissolved by another fungus or thoroughly chewed on by slugs and grubs. A successful oyster hunt often happens while looking for something else, but they can fruit for most of the warm season and prefer their wood to get wet.

If you want to introduce your family to wild mushrooms without risk, oyster mushroom growing kits make a fun project. You will receive a block of what looks like sawdust covered with white hair. Most companies use sterilized recycled coffee grounds since oysters are less picky than most. The white hairs are fungus mycelium, sleeping in its dried-out state till you soak it. The tiny emerging mushrooms will astonish you with the speed of their growth.

13. A FEW MUSHROOMS NOT TO EAT

Only a few mushrooms cause severe illness or death. Most of these belong among the death caps, a group in the genus Amanita. They contain a toxin that destroys the liver over several horrible days. Read everything in every guidebook you can find about these mushrooms, and never harvest anything you have any doubts about.

While morels are a prized early spring treat, another mushroom comes up simultaneously and can be mistaken for a morel. Honestly, I am not sure how people mistake it for a morel, which has honeycomb-like cavities, while the false morel looks like an orange lump with a stem. Some people consider these edible, but they can cause serious harm to others. Most guides do not recommend them.

14. IDENTIFICATION TRICKS: GILLS

When attempting to identify a mushroom, gather some key information. Color may not help since a few days can change a bright red mushroom into a yellowish mush. One factor that never changes: a mushroom either has gills, or it doesn't.

Whether a mushroom has gills or not has nothing to do with edibility. However, before you eat anything from your mushroom hunting, you need to identify it. Easy to recognize, gills make a great start to identification by eliminating lots of possibilities without them. Note the color of the gills and how they are attached to the stem.

15. IDENTIFICATION TRICKS: SPORE PRINTS

This time-consuming method of identification applies to people who want to eat some hard-to-identify mushrooms or who like to be very sure of their identification. The mushroom cap can be set on a piece of paper and allowed to sit in a room with no breeze overnight.

By morning, you should have a paper pattern that mimics the pattern of gills on the mushroom cap. The color of these spores can be key to identifying certain mushrooms. The prints are also pretty and can be preserved by spraying hair spray above the paper and letting it settle. If you cannot get a spore print on white paper, try dark-colored paper since some spores are white or pink. These also make attractive works of art for the mushroom lover, and you can make arrangements on the paper to create patterns. Just try not to sneeze.

16. HELPFUL TOOLS TO TAKE IN THE WOODS

All hobbies have tools, and mushroom hunting is no different. Every hunter has a preference, so please consider my choices a bit biased.

You need something in which to carry mushrooms. Options for this range from baskets to lunch boxes, but never carry or store mushrooms in a plastic bag. Plastic encourages moisture, and that allows bacteria to grow on the mushrooms. Keep mushrooms dry and with good airflow around them. Baskets work for this, but my favorite collecting container is a plastic shower caddy with a handle and sections for your various shower items. This lets you separate mushrooms till you can give them a better look at home.

If you need to identify mushrooms, make sure you do not crush or break them too much during your excursion. When I go to my chanterelle spot, where I know what I came for and will usually find lots of them, I take one or two mesh laundry bags. These hold a lot of mushrooms and allow good airflow. They will smash your mushrooms a bit, so only use this if you feel certain of your identification.

A magnifying glass makes a great tool for getting a better look at certain features. I have a friend who takes a dentist's small round mirror on a handle, which he uses to look under mushroom caps without upending them. A pocketknife, an indispensable tool when outdoors, will help you cut the edible parts off a tough shelf fungus, but never cut the stem of a mushroom off at ground level (see the section on identifying dangerous mushrooms).

17. MY PERSONAL FAVORITE MUSHROOM HUNTING TOOL

There is a tool for everything, including a special knife for mushroom hunting. I adore mine, and since I received it as a gift, I never go hunting without it. You can order these online, and most should be similar. You will not need a Swiss Army knife for mushroom hunting, just a few basic tools. My mushroom hunting knife has one blade, a flattened end to push away dirt, and a brush with stiff bristles.

If your knife lacks a brush, find a similar one. Tossing one mushroom covered with dirt into your container will get everything dirty, and most wild mushrooms turn into a mess when washed. Brush the dirt and bugs off before anything goes in with the others. If you ever must clean a bag of dirt-covered wild mushrooms, you will use more caution getting it off in the field next time.

Do not cut the stem of a mushroom to collect it unless you know what it is. The most lethal mushrooms in the world have an underground structure called a volvulus. If you miss noticing that, you could mistake these dangerous mushrooms for a harmless ones. Always dig around the base of the stem to make sure.

18. HOW TO LOOK

Mushroom hunting demands patience and attention. Prepare to spend hours walking through the woods, eyes fixed on the ground around you. Not only will you find mushrooms this way, but you will learn more than you can imagine about the woods around you.

Check old stumps and dead or dying trees. Walk all the way around; I have almost missed beautiful oysters on the other side of a tree. Try to identify everything, even the weird crusty ones growing on old logs. This builds your experience using your field guides, and it is also fun.

The day after rain can be a great time for mushroom hunting, but no reason to wait for the rain to go hunting. The forest floor stays damp for much of the year, so mushrooms pop up in dry weather too. If you see one, there may be more, so browse around and turn over leaves to see if any surprises are hiding from you.

19. WHERE TO LOOK (INCLUDING PUBLIC PROPERTY)

I find most of my mushrooms in the woods, but lawns can sprout crops of fungi too. Fairy rings of white mushrooms often emerge in a circle that gets slightly bigger every year as the underground mycelium eats everything inside it.

Check the usual suspects: dead or dying trees, open areas of forest floor, and around the base of big, mature trees (fungi usually specialize in conifers or hardwoods, but few can do both). I never stop scanning for something orange, and I have charged off looking for chicken of the woods, only to find a pile of orange leaves.

Sometimes you might notice a beautiful bunch of mushrooms growing around a tree in your neighbor's yard. How to handle this depends on your relationship with your neighbor. Knock and ask if you may harvest that mushroom over there, and people often allow you to. Do not sneak into any private or public place in the middle of the night to nab a nice specimen.

Rules for harvesting in public places vary by state and country. In most parks in the USA, you may hunt mushrooms for your own personal eating, but rules apply to collecting for profit. Delicate ecosystems may have rules not to pick or handle anything. Follow all rules on public property, and if you need more information, contact the property authority.

20. TIPS FOR PREPARING YOUR MUSHROOMS AT HOME

If you've followed the earlier advice, you kept your mushrooms in a container with some airflow on the way home. Even though many people think of mushrooms as slimy, most feel dry unless rained on that day or starting to get old. Do not eat a slimy, bad-smelling mushroom even if you identified it. Bacteria can turn a nice mushroom into a gross mess and can also make you sick.

Break up clumps of mushrooms as soon as you get home. Insects, grubs, slugs, and other crawly things love mushrooms, so expect to find that your dinner-to-be has a few extra residents. Also, breaking the clumps apart improves airflow and decreases moisture.

Recommendations differ, but no matter what someone else tells you, do not wash your mushrooms in the sink under lots of water! Most mushrooms absorb water, and many of us have watched good mushroom treats disintegrate into wet sponges. If your mushrooms are dirty, brush them off. I have a soft bristle toothbrush and a basting brush for this purpose. Once you have removed all the large debris, either wipe the mushrooms down with a damp paper towel or give them a very, very quick spritz with water.

21. THINGS YOU WILL LEARN TO PUT UP WITH IF YOU EAT WILD MUSHROOMS

First, you will end up eating some dirt. You cannot scrub a wild mushroom like a dirty potato, so getting every speck of dirt off can be a challenge. Chanterelles are the worst for this in my personal collecting experience since their cone-shaped caps appear designed to catch everything. Like mother said, a little dirt never killed anyone, so just clean them as well as you can and enjoy.

Second, you will probably eat or almost eat a bug. Mushroom hunters joke about this as "a little extra protein." Mushrooms attract many different things that want to eat them, from fruit flies to deer. Since both like the day after a hard rain, mushrooms and slugs seem to appear together, and slugs have eaten more of my prized mushrooms than I can count. If you munch a cooked grub, you will never know it, but I advise removing all slugs.

Third, you will hear "it's your funeral" or many other optimistic statements about your hobby. Ignore them. Most people assume that all wild mushrooms are poisonous and disgusting. Only a very few mushrooms will kill you, and learning to identify those is critical. People often post pictures in mushroom identification groups where they cradle a mushroom like it may be a grenade, wearing surgical plastic gloves and all.

For reference, you do not need to wear gloves to handle mushrooms. None of them have any toxin that could be absorbed through the skin. Wash your dirty hands when you are done, and all is well.

22. BASIC COOKING METHODS

With the growing popularity and availability of these exotic mushrooms, the internet has many recipes waiting for you. Look up recipes for your harvest: lion's mane excels as a crab or lobster replacement, and chicken of the woods can be battered and pan-fried. Puffballs can even be used as mini-pizza crusts.

Sauteing mushrooms brings out the flavor. Favorite recipes vary, but most involve butter or oil and a bit of garlic and herbs. You can roast or grill larger, sturdier mushrooms brushed with a similar blend. Wild mushrooms can be used in almost any dish that calls for ordinary mushrooms. Those with a strong flavor stand up to more seasoned dishes, but those with a more delicate flavor will disappear if there is too much else going on.

Assume that the pile of mushrooms you have on your counter will shrink while cooking. Just don't be surprised when they shrink a lot more than you expected. Mushrooms have a lot of water in them, and cooking this off leaves you with a much smaller mushroom pile.

23. TIPS FOR COOKING WITH WILD MUSHROOMS

When sauteing wild mushrooms, use whatever method you would use for store-bought mushrooms, but keep a close eye on them. Mushrooms contain a lot of water, and they shrink during cooking. As soon as they release their stored water, they can start to burn.

Use a regular pan, not nonstick, when sauteing mushrooms. They will not brown in a nonstick pan, and you lose a lot of flavor and texture. The butter or oil will absorb some of the flavors of the mushrooms. This works especially well for flavorful but very fragile ones like black trumpets.

If fortunate enough to find a big puffball, cut it into slices like burger patties. You can top these like a little pizza or bread and pan-fry them. You can probably imagine endless variations on toppings. Use them before they start to discolor, however.

Morels and chanterelles often go into sauces that highlight the more subtle flavor of these mushrooms. In contrast, lion's mane might go into a faux crab cake. Wild mushrooms have unique flavors, so look for a recipe to take advantage of them.

24. GIVING YOUR FIRST MEAL A TEST RUN

Some people can be allergic to a mushroom just like many other foods. Some people may not get along well with a particular mushroom. A few got upset stomachs in a group of people who ate chicken of the woods growing on a conifer. Most did not (it seems to be only chicken of the woods growing on conifers that have this effect).

If possible, cook up a small slice of the mushroom in a pan, then give it a taste. If you feel fine hours later, your body apparently likes that mushroom.

When serving wild mushroom dishes to others, let them know what you cooked and already tested it on yourself. If anyone feels uncomfortable eating wild mushrooms, do not make jokes or try to convince them. Eating wild mushrooms is not a zero-risk activity, and nobody should be bullied into it.

25. POSSIBLE MUSHROOM SIDE EFFECTS

If you have trouble with a wild mushroom meal, you may get an upset stomach. This can happen any time a person tries new food. If the upset stomach causes severe pain, lasts a long time, or otherwise causes concern, proceed to the nearest medical facility.

Any person trying a new food may experience an allergic reaction. While unlikely, this reaction can be dangerous, so seek medical treatment if you react to any food with hives, swelling, numbness in your face, or difficulty breathing.

If you have misidentified a mushroom, the most likely side effects include an upset stomach, nausea, and vomiting, or a meal that tastes terrible. Many non-edible mushrooms taste very bitter, and one can ruin a whole meal.

The primary exception, the death caps, should never accidentally end up in your food because you should never accidentally harvest one with the resources available. The death cap's side effects start off with the usual stomach problems, then a few days of peace and quiet before the toxins kill enough of your liver to cause liver failure.

26. STORING MUSHROOMS AT HOME

We often want to keep some around for the offseason with so many delicious opportunities and such a short mushroom hunting season. My chanterelle spot or a good-sized chicken of the woods haul will give up more than I can eat before they go bad.

The method depends upon the mushroom, and looking up the specifics for your samples will give the best results.

Dehydrating suits some wild mushrooms well. Due to black trumpets' delicacy, they make a wonderful seasoning if dried in a dehydrator or the oven and stored in a jar. Other mushrooms can be stored this way but get them very dry and keep them away from light and moisture. Soaking in water will rehydrate them.

Freezing works well, especially for thick, wet mushrooms like chicken of the woods that would not dehydrate well. Never try to freeze fresh mushrooms.

They contain so much water that it freezes and bursts the cells, leaving you with just a gooey remnant. The solution: sauté them in a regular pan (not nonstick) until they have released most of their water. They will then freeze for up to six months. A mushroom hunting friend uses ice cube trays so he can measure how much he adds.

In some Mediterranean countries, preserving mushrooms in oil is a popular method, usually after drying a bit. A fellow mushroom hunter tried a recipe for pickling chanterelles but found the results unpleasant. Still, you can certainly give this a try.

27. WHO SHOULDN'T EAT WILD MUSHROOMS?

Many experts recommend that you not feed wild mushrooms to children. This recommendation accounts for the smaller body size of a child, which would cause them to ingest more mushrooms per body weight than an adult.

Mushroom hunting families often disregard this recommendation, and some grew up eating similar mushrooms. In much of the world, wild mushrooms form an important part of the food supply. An associate from the Czech Republic learned mushroom hunting as a child, as did all her friends since wild mushroom hunting is a summer highlight in this culture. She finds it annoying that chicken of the woods does not grow there since I keep bragging about it.

Never put your child's life at risk just to eat something you found in the woods, but keep in mind that children all over the world do eat well-identified mushrooms without harm.

Anyone with a compromised immune system or elderly folks also should use caution eating wild mushrooms. Ensure everybody at the meal has a taste first to make sure they do not have an immediate bad reaction.

28. CAN WILD MUSHROOMS KILL YOU, AND HOW?

The wild mushrooms that kill people come from the genus Amanita. Not all of them cause harm, but only very experienced mushroom hunters who know how to identify these without mistake should take any risks with them. The Death Cap and Destroying Angel are some common names for these.

The toxins in these mushrooms target the liver. While the person will recover from the initial stomach trouble, the toxin continues to kill liver cells over the next few days. The liver, a large organ, takes a lot of damage before showing signs of liver failure. Still, by that point, the only solution is often a liver transplant.

The key to avoiding this fate: identify your mushrooms with care. Identification is not a joke, and being lazy can be fatal.

29. DANGEROUS WILD MUSHROOMS AND HOW TO SPOT THEM

The few mushrooms that can do serious harm are usually the Amanita genus. These, including Death Caps, keep coming up in discussion. That is because they will kill you, and identifying them is no joke.

Members of this genus are often (not always) tall, white mushrooms with gray or green on the cap. One species has a bright red cap flecked with little white bits, but that one has other effects we will discuss later.

These deadly mushrooms have an underground cup, a volvulus, that they grew up tucked inside like an egg. If you remove the dirt from around the base, you will often be able to find it. You can also find a ring around the stem where the top of the "egg" tore free as the mushroom grew. Both features can be hard to see on older ones.

They usually have white gills, but so do others. Check your field guide for a picture of how these mushrooms' gills attach to the stems since this can be an identifying feature.

In fact, memorize the basics of all your local dangerous mushrooms in your guides, never take any risks with a "maybe," and when in doubt, do what I do and only pick things when you feel sure of your identification.

30. LOOK-ALIKE, NOT TASTE ALIKE

Many harmless, tasty mushrooms have look-alikes, sometimes related and sometimes just by coincidence. Separating the good and bad can be easy if the mushroom has no common dupes (chicken of the woods) or tricky (most boletes).

Puffballs can make delicious meals, and some are bigger than your head. They are edible, but you must cut them open first. A real puffball that is not too old will be pure white and uniform all the way through. If it is discolored inside or has any structures that look like a developing mushroom, do not eat it. Many mushrooms have a veil over them when they emerge and can look like a puffball. If it is black on the inside, this is a pigskin puffball, a non-edible that usually grows on wood.

Chanterelles make delicious meals, and you can often collect a lot of them at one time. Jack o' lanterns come up at a similar time of year. While some of us puzzle over how you could mistake the two, it happens regularly. Cinnabar chanterelles, smaller and more orange than the usual ones, have a similar color. However, jacks have true caps, rounded on top and with gills underneath. Chanterelles have a flattened to indented top and false gills. Jacks will not kill you if you eat them, but your digestive system will reprimand you for it. One major difference: jacks always grow on wood. Chanterelles grow from under the soil, never on wood. Jacks grow in big, neon orange clusters, almost always around the base of a tree. To be sure, check for rotting wood underneath the surface.

Another group of mushrooms people collect, the genus Russula, has good and bad actors, but the two look very much alike. Low to the ground with flattened red or sometimes greenish caps, some of these mushrooms taste pleasant. Still, some have a bitter flavor, and some got the nickname "vomiter" for a reason. If your identification skills have not developed enough to tell the difference, or you just don't want them that much, feel free to leave this group alone.

31. DO PEOPLE COLLECT NON-EDIBLE WILD MUSHROOMS?

While much of the Western world has treated mushrooms either as part of their diet or as icky symbols of decay (especially during the melodramatic Victorian Age), Asian cultures have made mushrooms a major part of their cuisine but also a prized medical resource.

In traditional Asian medicine and modern natural health recipes, mushrooms nobody wants to eat become medicine by making tinctures or powders. Chaga, a birch tree parasite, looks like a blackened lump of charcoal plastered to a tree, and nobody collects it for dinner, but medicinal mushroom hunters prize it.

Reishi also finds a place in the mushroom medicine cabinet. Among the most well-known medicinal mushroom, it comes in several species native to Asia and North America. The mushroom forms a hard, flat shelf (or sometimes many shelves), and the North American variety prefers hemlock. Too hard to eat, it makes its way into different forms of traditional medicines.

Some medicinal mushrooms also go into cuisine. Both shiitake and maitake (hen of the woods) make good meals. Still, traditional medicine reports immune-boosting, cancer-killing, and other properties. Shiitake, grown in cultivation, can be found in upscale grocery stores, but maitake only grows wild.

Many other mushrooms have traditional medicinal uses. One that caught attention in the recent news: cordyceps, a group of fungi that parasitize, mind-control, and then kill and burst out of their host insects. In China, a particular species of this fungus infects caterpillars. The resulting spore-bearing yellow spikes become the most expensive fungal fruiting body on the planet. Its reported powers include improving sexual and athletic performance.

The latter claim drew attention after Chinese Olympic athletes seemed to perform unexpectedly well. After accusations of doping, they reported only using this caterpillar fungus, not a banned substance. How much this mushroom affects the body has not been determined.

All untested medicinal claims for anything should be treated with caution and never used without consulting your doctor. Biologically active materials in these fungi could react with medications you already take, among other concerns.

32. WHAT HEALTH BENEFITS MIGHT MUSHROOMS HAVE?

Traditional Asian medicine reports many benefits from medicinal mushrooms. Scientists continue to test them to determine if medicinal compounds can be isolated from them. Since mushrooms have thousands of mating strains instead of "male" and "female," the genetics become complicated.

Some of the benefits listed for medicinal mushrooms (and this list includes properties ascribed to some or all of those listed above) include curing cancer, improving stamina and performance, improving digestion, treating diabetes and high blood pressure, improving the immune system, act as antibiotics, kill viruses, prevent dementia and Alzheimer's, cure liver disease, improve memory and cognitive performance, lower cholesterol, prevent heart disease, encourage weight loss, enhance fertility, prevent cavities, and eliminate allergies.

While unlikely that any mushroom can do all or even most of these things, scientists have found that some of these mushrooms have interesting biological activity and might one day supplement standard treatments. Again, do not use these without your doctor's approval. If your doctor gives the OK, you can add some of these mushrooms (shiitake) to your regular diet or go hunting others if you are in their territory.

As people seek alternatives to modern medicine and its side effects, traditional remedies become more popular. Use caution since a medically active substance, in a pill or a mushroom, can do harm or good depending on how you use it.

33. WHERE CAN YOU FIND MEDICINAL MUSHROOMS?

If you live in North America, maitake or hen of the woods makes a consistent appearance in early autumn. If you find a tree hosting one of these large fruiting bodies, take note and come back next year. While some prize it for medicinal purposes, others just like to eat it. Please see the section above about "hen of the woods," aka sheep's head or ram's head, for information about hunting them.

The North American species of reishi grows on hemlock, so if you wish to discover it, go searching in hemlock or mixed hemlock forests. My yearly sighting occurs on one dead hemlock tree in one of my hunting areas. Although I do not harvest this mushroom, the shelves in early summer look like deep mahogany wood with ten shellac layers on top and a white border. Most fall off over the winter, but last year's shelves lack the shiny surface or the white, still-growing border.

If you wish to try hunting cordyceps fungi, you will have to find some infected insects and wait for the fungus to explode out of their bodies. I have found a few parasitized dead insects. Still, this mushroom's high cost reflects the fact that each fruiting body reaches only a few centimeters at most. The species used in China only affects a certain caterpillar species, so it seems unlikely you will collect any unless you live in that region. Videos of this fungus in action will give you a disturbing and/or fascinating evening of YouTube watching.

34. HOW DO OTHER CULTURES USE WILD MUSHROOMS?

We discussed the cultural use of medicinal mushrooms in Asia. Still, mushrooms also form a major part of many Asian dishes. Many edible North American mushrooms have counterparts in Asia, although they often like different trees or locations. The American willingness to try anything other than a commercial button mushroom can be attributed in part to exposure to traditional and modern Asian cuisine.

In Europe, Slavic and Baltic countries have an especially long and close relationship with mushrooms as a part of their diet. Mushroom hunting continues to be a family activity in many of these countries, with children learning from parents and grandparents how to determine which are safe. With this extensive in-field training from early childhood, my friend from the Czech Republic identifies things at a glance and eats mushrooms I did not know were edible. In these countries and parts of Russia, where winters can be harsh, dried mushrooms have long been a way to flavor and add nutrition during long, dark days. Even bitter boletes, soaked in repeated water baths to leach out the bitterness, get eaten or dried for future use.

Mexico and South America have their own preferred mushroom dishes. In Mexico, a fungus known as corn smut infects ears of corn, turning the kernels into what many people see as a stinking gray mold. However, this mold-altered corn is a Mexican delicacy. In many traditional cultures of these regions, psilocybe mushrooms, also known as "magic mushrooms," have been used for centuries. Shamans or medicine men of many cultures have used hallucinogenic plants and mushrooms to enhance their visions and abilities.

Mushroom hunters looking for a "magic mushroom" trip should keep in mind that identifying these tiny brown mushrooms from among thousands of other species of tiny brown mushrooms should be left to experts, as some misidentified ones will just make you ill.

35. CAN YOU HARVEST MUSHROOMS AND MAKE YOUR OWN MEDICINE?

You can harvest any medicinal mushrooms that grow in your area. If you do, research the best ways to prepare them for medicinal use. Many do not grow in the form of traditional mushrooms and would not go well in a dinner dish. Hard fungi, like shelves and brackets, need processing before use.

Tinctures involve extracting the active contents of the fungus in alcohol. They can also be dried and the whole mushroom used as a powder. Some, like shiitake, make their way as fresh mushrooms into cuisine. Many believe that eating these fungi fresh provides the greatest benefits. Fresh reishi, a hard shelf, can be brewed into a tea. With the bitterness disguised by other flavors, it has great value as traditional medicine.

36. HOW LONG HAVE PEOPLE BEEN HARVESTING MUSHROOMS?

The Iceman, a Copper Age mummy discovered in Europe, dates to 3000 to 3500 BC. Analysis of the body indicates that he suffered from intestinal parasites, like many of his contemporaries. However, this traveler carried a remedy with him: birch polypore, an edible but not at all tasty shelf fungus. People may have chewed on these as a worm control medicine.

The Iceman also carried several kinds of mushrooms that, when dried, make excellent fire starters. Properties of mushrooms as medicine, food, and tools had achieved common use long before the Iceman. Because mushrooms are so delicate, research rarely turns up evidence of an ancient fungus. However, mushrooms feature extensively in the mythology of many ancient cultures.

37. MUSHROOM HUNTING: OLD METHODS WORK

You can download a dozen mushroom identification apps for your phone, but none of them offer any accuracy. Besides a few one-of-a-kind standouts, most mushrooms look much alike, and a single picture has no information to help with a correct ID.

A field guide may require a bit of reading, and some take time to learn to navigate. However, they will teach you how to sort out the differences between look-alikes and relatives. Many excellent books can teach you about the broader place of mushrooms in our ecosystems, making the mushroom hunting experience more enjoyable.

While a few mushrooms grow in captivity, most will not, and probably never will. Morels, chanterelles, and others that form long-term intertwined relationships with partner trees will not grow without this partnership, which may take decades to develop. Since you will probably never find these special treats in a store, learning to find them in the wild requires the same skills it always has: patience, experience, and luck.

38. MOST IMPORTANT SKILLS TO LEARN FOR MUSHROOM HUNTING

If you want to find wild mushrooms, go on lots of walks in the woods or other wild places. Even if you see no fungi, get used to noticing as many things as you can. Notice whether you see pine needles or oak leaves on the ground. Take note of areas with large mature trees.

Intuition tends to be the brain's way of telling you to stop ignoring interesting things. If something tells you to walk through brambles to get to that stump, listen to it. You might find nothing; you might find delicious oysters. If you never leave the path, you might not find anything.

My chanterelles taught me that impatience gets you nowhere with mushrooms. Even the rare few that pop up like they had an alarm set can be no-shows under less-than-optimal conditions. If a likely area has nothing, come back in a week. Most mushrooms vanish in a few days.

Whether you believe in luck or coincidence or other forces, mushroom hunting requires a dose of something to lead you to those great finds. I prefer to consider it intuition. Regardless, do not ignore it. I have taken a path I had not planned on and brought back a shirt full of mushrooms.

39. NEVER GUESS ABOUT THESE THINGS WITH A MUSHROOM

You should never guess about anything when dealing with wild mushrooms. Make sure your identification doesn't lead you astray. Guidebooks can only do so much, and as mushrooms age, they do not look like the guidebook photos anymore.

If these rules seem random, read the section on Amanita mushrooms in a guidebook or online.

Never pick a mushroom and assume it has no volvulus (the underground cup). Always brush the soil away and look. Never assume that no volvulus or no ring around the stem means this mushroom is safe. A mature Amanita can lose those features as it ages.

Never guess if you feel unsure of your identification. Mushroom groups online can assist, but most of them are not professionals. If you are guessing, you may experience a very unhappy stomach or possibly worse.

40. REASONS TO TAKE KIDS MUSHROOM HUNTING WITH YOU

Kids benefit from time outside, learning about nature from adults in their lives. Taking children mushroom hunting with you can open them up to a whole new world of learning. Because current generations spend more time indoors and looking at screens, they benefit from developing observation skills in the real world.

Some kids may find these excursions boring. You may, however, plant the mushroom hunting bug in some young person's head. The hobby becomes addictive, costs almost nothing to pursue, and gets people active and outdoors.

Learning outdoor skills can help kids develop a sense of mastery. People feel that they have made great progress in a skill have more confidence along with more knowledge. Your kid's classmates might not be impressed with his or her new hobby, but minds may change when they see a picture of said classmate hauling a 30 lb hen of the woods back from a trip.

41. HUNTING MUSHROOMS AND RESPECTING NATURE

To keep your mushroom hunting nature-friendly, take some general precautions to keep your footprint small. You should not need to be reminded of obvious things: do not leave trash or anything else behind. Collecting something to learn about it may be fine but kicking things over or wrecking them for no reason is unacceptable.

Limit your mushroom hunting in wildlife preserves or other special nature havens. Even if harvesting in ordinary parks or game lands is legal, it may be illegal or harmful to take things out of these areas. They may contain a delicate protected ecosystem. Unless you have specific permission, consider these places "do not disturb."

While reaching a tricky mushroom can require some navigating, try not to trample the plants, or turn up the soil more than necessary. You are not destroying the entire forest, but more mushroom hunters come behind you, and the cumulative damage adds up. Mushroom hunting may require you to leave the well-trodden path but try not to tread on too much other wildlife while searching.

Also, keep in mind that if you have your phone, you have a camera. If you see an interesting plant or mushroom you do not need to disturb, you can always take pictures from various angles instead. This can make great practice for using your field guides, too.

42. WHY WE'RE NOT TALKING ABOUT HALLUCINOGENS IN THIS GUIDE

Humans have known about hallucinogens and other drugs for most of human history. While yeasts are fungi and without them, we would not enjoy alcoholic beverages, most people talking about the genus Psilocybe's hallucinogenic mushrooms. These little brown mushrooms grow worldwide, but the best known and one of the most common grows throughout Central and South America. Prized by shamans, they were used for spiritual journeys, not recreation.

Today, medical research probes the possibility that these mushrooms might have applications in psychiatric treatment. However, few people recommend testing on yourself. These nondescript little caps get mistaken for hundreds of other little brown species with nothing interesting to offer, or possibly a bad stomachache.

Because the chance of your little brown mushroom discovery being a hallucinogen is slim to none, I recommend you stick to edible mushroom hunting and leave the "special" ones for the professionals.

43. ONE MUSHROOM TO AVOID UNLESS YOU WANT TO TAKE A "TRIP"

Have you seen pictures of fairies sitting on a mushroom with a bright red, white-flecked cap and white stem? This is fly agaric, a member of the Amanita genus containing the most lethal mushrooms in the world.

Unlike the little brown hallucinogenics, fly agaric does not hide. These sturdy mushrooms prefer to grow under pines, and their bright caps catch the eye. In parts of the USA, these mushrooms have red caps, but caps are yellow or orange in the northern part of the country. The flecks of white are remains of the veil that covered the mushroom inside its "egg."

The remains of that "egg" are the volvulus at the base of the stem.

The long history of these mushrooms in fairy tales does not just come from their appearance. They contain a potent hallucinogen, but the trip will not resemble the more laid-back ones induced by psilocybin. Warriors have used the mushrooms to induce a "berserker" state of hyper-energized rage. Few have described it as pleasant. You will see this one on your mushroom hunting but leave it alone.

44. MUSHROOM HUNTING COMPETITION TO WATCH OUT FOR

Mushrooms contain many vitamins and minerals, so non-human mushroom hunters also pursue them when they can. While you may not be able to do much about it, you can look for signs of who got your mushrooms before you did.

Slugs, first of all, love mushrooms. The bane of mushroom hunters, they manage to go from slug speed to teleportation if they hear a fresh oyster mushroom has popped up. Beetles and flies also lay eggs and let their larvae enjoy the meal.

In my area, white-tailed deer and several species of squirrels love mushrooms. If deer have been at work, I can usually spot hoofprints. Squirrels, especially American red squirrels, usually leave the stem. They also prefer pines and will be found screaming at you from above. If you ever find a mushroom tucked neatly into the branches of a tree, assume a squirrel has been at work.

If you find that someone rather than something is getting to your spots, you have two options. One is to meet these folks and see if you can share or divide the hunting, and the other is to get up earlier than they do.

45. SIGNS THERE MIGHT BE MUSHROOMS AROUND

How do you know if you are in an area to find mushrooms? If you have a guidebook specific to your region, it should help guide you. Another technique: consider where familiar mushrooms in your home area grow and check those types of areas. Never get lazy with identification. Several Laotian immigrants died of Amanita poisoning after mistaking them for an edible mushroom in their home country.

Mushrooms and trees have a deep relationship, but fungi predate trees by a few hundred million years. Open fields, especially those housing large animals and their droppings, provide an opportunity to find some remarkably interesting if not edible mushrooms. Take an artillery fungus home from a pile of cow manure, and you can watch it launch a spore packet across the room.

Mushrooms produce no chlorophyll, so they have no preference for dim or bright light. However, they do prefer moist areas, and bright sunlight tends to dry out the soil.

The best way to predict if you will find mushrooms? Know where they came up last year. Nobody can guarantee a repeat, but you would be wasting an opportunity if you did not check successful past spots.

46. AM I HURTING THE FUNGUS IF I PICK ALL THE MUSHROOMS?

If you pick a plant, it dies. If you pick apples off a tree, however, the tree remains. The apples carry the tree's seeds, and apples taste good because the tree wants animals to eat its apples and carry the seeds away. Fruits and berries exist to lure animals into eating them.

Mushrooms act more like fruit on an apple tree. The fungus itself lives as a thick, tangled mat of mycelium that can cover a large amount of underground real estate. It can spend decades spreading and growing before it ever makes a mushroom. The fungus does not really "want" you to pick its mushrooms, but they last such a short time anyway that you do no harm to the fungus body.

Some debate exists about whether over-harvesting, like the massive morel harvests in the Pacific Northwest where professionals arrive and sweep the forest bare of anything they can sell, harms mushrooms. No proof exists either way, but these commercial harvesters, with their total disregard for the environment, do nothing to improve mushroom hunters' reputation overall.

47. WHY HAS MUSHROOM HUNTING BECOME MORE POPULAR?

In many parts of the world, mushroom hunting never lost popularity. In parts of Europe and in the United States, mushrooms (or "toadstools") became symbols of rot, decay, foulness, and disease. Of course, they do a good bit of decaying, but if dead material sat around untouched, it might soon become a problem.

Many restaurants now serve "fusion" cuisine incorporating flavors from other cultures into their food. This has brought many wild mushrooms closer to mainstream recognition. Vegetarians also like experimenting with mushroom flavors since they can add a pleasant meatiness without the meat.

In a culture of store-bought everything, mushroom hunting offers the excitement of finding things other people will never taste, that you found through your persistence and patience and that you used nothing but your own skills to discover.

Survivalists, serious or amateur, also add a new demographic to people interested in eating what the land provides, and mushrooms can provide food and critical nutrients during hard times. Taking advantage of all available resources would give you an advantage in a survival situation. People who never considered mushrooms as a serious food item before now find that besides the health benefits, they also preserve well and add great flavor and nutrition to your survivalist meal.

48. LOCAL CLUBS AND GROUPS

As mushroom hunting grows in popularity, clubs and groups have popped up in many areas. I pay $15 per year in membership dues to be part of our local club, which has mushroom walks during the season and has speakers and classes during colder weather. The group also participates several times a year in a biodiversity challenge held by a local nonprofit conservation group, collecting lists, and photographs of edible and non-edible fungi to help with their surveys.

Local groups often have resources about the mushrooms in your area and their unique habits. They can recommend good guidebooks that cover your area. Some have photo competitions and other challenges. Many will have veteran mushroom hunters who lead hunts throughout the mushroom season. Help from an experienced guide can improve your identification skills and keep you safer. The knowledge gained from years of experience can be combined with a good guide to help keep your identification skills sharp.

49. WHY PEOPLE THINK MUSHROOMS ARE BAD

The view of all fungi as bad, slimy, gross, or dangerous emerges most in parts of the world where people have not harvested or collected their own food for a long time. Until the recent resurgence of interest, most people who spend almost no time exploring outdoors only see a fungus when it molds their leftovers or in a jar of sliced button mushrooms.

Because fungi do not belong to the plants or the animals, they look and behave unlike our other food. While people have trusted the commercial button mushroom, many people still refuse to touch a pizza or any dish with a few mushrooms in it. With their strange shapes and activities, they challenge non-adventurous eaters.

The common belief that all or most mushrooms are poisonous is incorrect but widespread. Many people on Facebook pages for mushroom identification (and if you use them, take all advice with a grain of salt) often post pictures where they will only touch the mushroom with rubber gloves on or will not touch them. Only a few mushrooms will cause harm, and only if you eat them. You cannot be poisoned by a toxic mushroom by touching and handling it. The toxins do not absorb through the skin. It is always safe to touch, investigate, and learn more about mushrooms you find. Wash your hands when you get home, but the presence of irritant plants should encourage that anyway.

50. WHAT MUSHROOMS CAN TEACH US ABOUT THE WORLD AROUND US

Humans interact every day with members of the plant and animal kingdoms. These organisms aren't more common than fungi, just more visible. While plants spread out seeking light, the entire body of a fungus can stay buried under the forest floor for decades or centuries, stopping only when the food supply runs out. In a large undisturbed forest, few factors may exist to limit the fungus' growth, and they can spread over massive areas.

Mushroom hunting can open our eyes to the entire world of fungi around us and all the critical things they do to keep our ecosystem working. A wood-decay fungus fruiting on a standing tree might look innocent, but it means the tree's death approaches. The fungus inside may not have killed it, but its digestive enzymes will break down some of the wood's tough cellulose before the tree even falls. Without fungi, the forest would bury itself in dead wood and organic debris.

Most people know that the first antibiotic, penicillin, came from a fungus. Most other common antibiotics also come from fungi. Why do fungi spend so much energy making drugs to kill bacteria? Bacteria and fungi compete for food, and nobody likes to lose. Fungi excel at kinds of chemistry humans have yet to master, and antibiotics work safely in humans because they attack things bacteria have that animal cells do not, like a cell wall.

While we have talked about mushrooms' edible and medicinal aspects, consider the huge impact they have had on cultures across the world.

Many books discuss the larger cultural, ecological, and social impacts of mushrooms on the world we live in.

HELPFUL RESOURCES:

National Audubon Society Field Guide to North American Mushrooms (National Audubon Society Field Guides) by Gary Lincoff, available from most booksellers.

The Complete Mushroom Hunter, Revised: Illustrated Guide to Foraging, Harvesting, and Enjoying Wild Mushrooms by Gary Lincoff, available from most booksellers.

Mushrooms of the Northeast: A Simple Guide to Common Mushrooms by Teresa Stallone and Walt Sturgeon, available from most booksellers (the series includes guides for other areas of the USA)

Wild Mushroom Hunting Forums (https://wildmushroomhunting.org/), not limited to the USA

Learn Your Land (https://learnyourland.com/), a blog and educational YouTube series by Adam Haritan, updated regularly

READ OTHER 50 THINGS TO KNOW BOOKS

50 Things to Know to Get Things Done Fast: Easy Tips for Success

50 Things to Know About Going Green: Simple Changes to Start Today

50 Things to Know to Live a Happy Life Series

50 Things to Know to Organize Your Life: A Quick Start Guide to Declutter, Organize, and Live Simply

50 Things to Know About Being a Minimalist: Downsize, Organize, and Live Your Life

50 Things to Know About Speed Cleaning: How to Tidy Your Home in Minutes

50 Things to Know About Choosing the Right Path in Life

50 Things to Know to Get Rid of Clutter in Your Life: Evaluate, Purge, and Enjoy Living

50 Things to Know About Journal Writing: Exploring Your Innermost Thoughts & Feelings

50 Things to Know

Stay up to date with new releases on Amazon:
https://amzn.to/2VPNGr7

50 Things to Know

We'd love to hear what you think about our content! Please leave your honest review of this book on Amazon and Goodreads. We appreciate your positive and constructive feedback. Thank you.

www.ingramcontent.com/pod-product-compliance
Lightning Source LLC
Chambersburg PA
CBHW070300220526
45465CB00004B/1687